CS Publishing, Seattle 2014
ISBN 978-1502940872
ASIN : B00NZ2RZRQ
Copyright, Lucio Giuliodori 2014.
www.luciogiuliodori.net

CONTENTS

1 From hermetic philosophy to quantum mechanics... p. 8

2. Experiments, paradoxes and theorems: an introduction to non-locality... p. 22

3 An interconnected universe: philosophical implications of new physics... p. 43

*Because mine is not a science like any other:
it can't be communicated in any way.*

Plato.

# 1. From hermetic philosophy to quantum mechanics.

The Italian musician Franco Battiato, in collaboration with philosopher Manlio Sgalambro, has recently written an opera inspired by the Italian Renaissance thinker Bernardino Telesio. The work was presented to the public in holographic form, designed a priori in the refined mind of the Sicilian artist, whose reflection is often fraught of perennial, esoteric philosophies.

«Witnesses at the representation in Cosenza say that the public believed that singers, dancers and the actor Giulio Brogi were on stage, in flesh and blood, deceived by the incredible effects of the holograms»[1]. Therefore, a sort of reality shift is what the audience experienced.

However, what exactly is a hologram? The celebrated American physicist David Bohm explains: «Relativity and quantum theory

---

[1] From the website of the Sicilian singer: http://www.battiato.it/?cat=6

imply a divided totality in which analysis of well-defined separate parts is no longer relevant. Nevertheless, there is a tool that can give us an immediate perception of the meaning of this unity [...]. Such a perception is possible when we consider the hologram (the name derives from the greek *holos*, meaning «all» or «whole», and *gram* which means «to write». A hologram is therefore a tool that, so to speak, «writes the entire»)»[2].
What many scientists now argue, carrying on the work begun by courageous David Bohm, is that the reality we experience every day, which seems real and concrete, is in fact nothing but a mere hologram. Among these

---

[2] D. BOHM, *Universo, mente, materia*, Como: Red edizioni, 1996, p. 200. My translation. The same applies to every non English written book quoted from now on.
Nevertheless, Plato in his *Timaeus*, had already asserted something very similar: «The seeds of all things were thrown into a crater and every soul contains all of them. The soul is like that crater; being a mixture of everything, the soul is able to recognise everything and because of this intimate affinity, it experiences a universal life. The soul can empathise with each power of nature, not through reasoning or violent efforts, however - Plato says - with a similar movement to the growing of the hair or teeth, with an almost unconscious development». E. ZOLLA, *Auree*, Venezia: Marsilio, 1995, p. 49.

various scientists, let me primarily refer to the neurophysiologist Karl Pribram who had collaborated with Bohm. The two scientists worked together, sharing the validity of the holographic model. Pribram proposed it for his studies of the brain, arguing that it could provide the most valid explanation to the classification of memories and therefore, de facto, to their non-locality[3].

---

[3] To this regard, concerning the affinity of certain aspects of quantum physics with the hermetic philosophy of the Renaissance, it would be interesting to evaluate a parallelism between this conception of memory and the famous techniques of Giordano Bruno. Those of Nolan were not only strengthening mnemonic techniques but real operational tools for transmutation of microcosmic and macrocosmic reality, i.e. changing the individual in order to transform the whole of reality: «Bruno not only intends to strengthen the mnemonic muscle, rather he wants to change the cosmos. That is - let me say again - change the very structure of every initiate's mind and then, through them, change the world, by means of the micro-macrocosm interdependence, in other words mind-universe. This is the work of «his» Ermeticism, changing man's mind, changing his intimate essence and then, through him - newborn-man, revolutionising the world». G. LA PORTA, *Giordano Bruno. Vita e avventure di un pericoloso maestro del pensiero*, Bologna: Bompiani, 2001, p. 195.

As Elémire Zolla stated in *Verità segrete esposte in evidenza*: «One cannot demonstrate a difference between the perception of reality and a constant and durable collective hallucination: they are the same thing»[4].
Unquestionably, this concept of reality seems shocking and absurd at first, even ridiculous. Yet, as if that wasn't enough, there is another side to the issue that is even more shocking and absurd, but far from ridiculous: this worldview is supported and validated by real experiments, years and years of laboratory research carried out by the greatest minds in contemporary physics, such as Nobel Prize winners Albert Einstein, Wolfgang Pauli, Max Planck (who devised the concept of the quantum of energy), Louis De Broglie, Richard P. Feynman, Werner Heisenberg, Paul Dirac, J. Von Neumann and obviously Stephen Hawking.
These are just a handful some of the many famous names associated with quantum mechanics.

---

I refer the reader to K PRIBRAM, *Languages of the Brain: Experimental Paradoxes and Principles of Neuropsychology*, Upper Saddle River: Prentice Hall, 1971.
[4] E. ZOLLA, *Verità segrete esposte in evidenza*, Venezia: Marsilio, 1990, p. 159.

Nevertheless, let us now go back to the opera and ask ourselves why Battiato has compared holograms, with obvious reference to David Bohm, creator of the holographic universe theory, with Italian Renaissance philosophy?

«Hermetic thought had established the interrelationship and interconnectedness of all things, so that if in any one area of the textile of reality a filament is pulled, in another point of reality a filament is stretched or slackened. Nuclear physics, at least, has confirmed the validity of this principle and has translated the theory into practice. [...] At the beginning of this century, biology, chemistry and physics were presented as three separate and distinct disciplines [...] When they were established, these disciplines were hailed as innovative and revolutionary, while they are nothing more than the expression of Agrippa and Paracelsus' global reality. They express a unity that existed long before analytical processes would come to create an artificial distinction between its components. In reality, biology, chemistry and physics have always been interconnected and Cartesian science has made a mistake assuming that they were separate»[5].

«All is One» then, but what does this mean *practically*? For Renaissance hermetic philosophers the universe was conceived as a large animated body, enlivened by a principle, the *anima mundi*, by dint of which everything was substantiated by occult and spiritual energy. In this perspective, subject and object exchange roles, as everything (animated by the *anima mundi*) is both object and subject of the same reality. The analogies with Bohm's physics are quite evident; let alone the implicated and explicated reality in Cusano (*complicatio* and *explicatio*), Bohm even uses the same words: implicit and explicit order.

We should consider that affirming the illusion of reality is anything but sensational news, as far as that statement sounds absurd and paradoxical, it is absolutely in line with what has always been claimed by various traditional thinkers. From Gnosticism to Hinduism, from the above mentioned Renaissance magicians to theosophists, up to the Twentieth Century with the most famous

---

[5] M. BAIGENT – R. LEIGH, , *L'elisir e la pietra. La grande storia della magia*, Milano: Il saggiatore, 2003, pp. 268-9.

traditionalists such as Schuon, Guénon, Evola and Cooramaswamy. The list would be even longer, however, what matters here is the thread that binds all these authors: all these currents of metaphysical thought, finds a shocking confirmation in the current discoveries of quantum mechanics and this is why, in light of the foregoing, speaking about "physics and metaphysics" should now sound overtly pertinent.

Macrocosm and microcosm interpenetrate and the individual is not just a walk-on actor but an effective, fundamental character of this interpenetration: he is the subject and object of this reality; being a fundamental and intimately integral part of it, as he himself changes, reality changes too. In a nutshell, he reminds us of the Renaissance *homo faber*, the *magus*, the master of his own destiny, the god-man mentioned by both Eastern and Western traditions:

«Through progressive emanations, divine ideas penetrate material forms by means of the *spirit mundi*, right down to the most minute particles. Nature is thus deified, as an essential complement to deity.

An Infinite God cannot be separated from the other infinite worlds of which he is the cause and animating force.
Man is an essential part of nature and the presence of the divine in it inevitably results in the presence of the divine in the human, manifesting itself in that heroic fury, relentless effort in itself, never tired, never satisfied with the search for truth»[6].

Francesco Pullia, in his essay on the affinity of Bruno's thought with Eastern philosophies and *de facto* the affinity of both with quantum mechanics, asserts:

«Man is not a separate subject from nature and the divine, rather he is, by his very essence, nature and divinity, he is no longer at the centre of the cosmos since everything is composed of the same living matter, the same energy: there is nothing of ours that becomes extraneous and nothing extraneous that doesn't become ours, says the philosopher. [...] One cannot but note, how this holistic interrelationship, unchanged for millennia, marks the main cognitive core of modernity's horizon from Einstein up to the contemporary

---

[6] G. LA PORTA, cit. p. 124.

outcomes of Fritiof Capra, Humberto Maturana, Francisco Varela, Gregory Bateson»[7].

What the ancients had intuited - or probably even experienced if we think of the role of magic in the Renaissance[8] - is now being demonstrated in the laboratory. This factor casts quantum mechanics to a top cultural position and imposes a rigorous study which will shortly revolutionise not only the way we do philosophy, but also the way we live our everyday life probably.
This revolution is deeply affecting the various branches of knowledge, once split by Cartesianism and all the materialistic conceptions of modernity that have divided a

---

[7] F. PULLIA, *Giordano Bruno tra Oriente e Occidente*, in "Testimone dell'infinito. Giordano Bruno 1600-2000", Atti del Convegno Perugia-Terni, Ali&No Editrice, Perugia 2004, p. 75.
[8] «This world of the second century was, however, seeking intensively for knowledge of reality, for an answer to its problems which normal education failed to give. It turned to other ways of seeking an answer: intuitive, mystical, magical». F. YATES, *Giordano Bruno and the Hermetic Tradition*, London: Routledge and Kegan Paul, 1964, p. 4.
See also M. DONA', *Magia e filosofia*, Milano: Bompiani, 2004.

reality that was already united: «The *forma mentis* that has shaped the modern world derives to a greater extent from Descartes. In the Western world, Cartesian thought was meant to set in motion a revolution as profound as the one created by the Renaissance, a revolution that would lead to the Enlightenment, also known as «The age of reason»[9]. (This is why, the hermetic vision of reality, during the rationalist «triumph» age,

---

[9] M. BAIGENT – R. LEIGH, cit., p. 247.
Psychology itself is not exempt from this fragmentation, that is why Assagioli, through his Psychosynthesis proposes a sort of recomposition. In addition to the main four forces - behaviourism, psychoanalysis, existential-humanistic psychology and transpersonal psychology - «in his writings Assagioli added a fifth force called *psychoenergetics*, a psychological consequence of the new scientific paradigm shift followed by the revolutionary significance of quantum physics' new discoveries. The father of Psychosynthesis believed these findings would set the way for a new vision both of the world and man, whose great truths – stated by all sacred traditions over thousands of years - were eventually accepted without prejudice. These truths would be finally given the importance they deserve, by dint of their valuable insights and directions for the future development of humanity». P. GUGGISBERG NOCELLI, *La via della psicosintesi*, Firenze: L'Uomo Edizioni, 2011, pp. 95,6.

found shelter in the arts; let us think of William Blake as an example).

As it completely affects the pragmatic sphere of our existence, the philosophy implied by the new physics claims a role of radical importance, probably almost never seen before.

Reminding Sgalambro's words: «A philosophy that does not make you forget all the others is not worth anything». We have probably found this philosophy today: it is the reflection that derives from the fundamental assumptions of quantum mechanics.

This short essay aims to present the outlines of some of the many philosophical implications arising from what might be called a true cultural revolution, the revolution of «quantum meta-physics»[10], as theorized and demonstrated by the famous physicist David Bohm.

---

[10] M. SGALAMBRO, *Anatol*, Milano: Adelphi, 1990, p. 22.

*Consciousness is the theater, and precisely the only theater on which everything that takes place in the Universe is represented, the vessel that contains everything, absolutely everything, and outside which nothing exists.*

Erwin Schrödinger

## Experiments, paradoxes and theorems: an introduction to non-locality.

In order to reach the targets this essay set, it's necessary to briefly introduce how and why the (dangerous) path undertaken by some brave physicists settled (let us think of Bell, Aspect, Pauli, Bohm and Pribram just to name a few), whose genius and open-mindedness, has brought about a determinant watershed in the development of quantum mechanics. Among these eminent scientists an important position certainly pertains to the courageous American physicist David Bohm.

According to Bohm, physics has always been concerned mainly with its formal aspect rather than its content, aiming mostly to solve the problems rather than *understanding* them (as well).

According to the bohemian conception of the world, truth is not only concerned with mathematics. As the physician himself states: «Today the essence of physics is considered essentially a mathematical one. You can feel that truth is in the formulas. In this day and age, my colleagues hope to find an algorithm which can explain a wide spectrum of experimental results, but it will still have inconsistencies»[11].

Although fundamental, the mathematical formalisation of physics is not enough if it is not supported by intuition and creativity[12].

David Bohm was not satisfied with a deterministic description of the universe as it was provided by Einsteinian relativity, in which stale Newtonian elements were still present. With his meta-physical yearning, Bohm implicitly pulled down dams and barriers that divided science and philosophy: the description and the construction of reality that he was looking for, and was gradually finding, was ineluctably philosophical, even «spiritual»[13].

---

[11] Quoted in M. TEODORANI, *David Bohm. La fisica dell'infinito*, Cesena: Macroedizioni 2006. My translation.

[12] See D. BOHM, *On creativity*, New York: Routledge, 1996.

[13] In addition, Bohm's cosmology was thorny in the eyes of the academic world that surrounded him. That world was prohibitive and discrediting, certainly not yet ready for evidence whose absurdity is more appropriate for literature than physics. In some ways, the path to the American scientist reminds us of Tesla's, as genius often, in addition to wonder, arouses envy and resentment.

From the scientist's own words:

«Particles must be conceived as certain types of abstractions within the total field, corresponding to regions where the field is very strong (known as «singularities»). As we move away from singularity, the field fades, until it merges imperceptibly with other singularities, but without fractures or divisions. Therefore, the classical idea of a world separated into distinct interacting parts is no longer valid or relevant. Instead, we must consider the universe as an undivided whole and without fractures. And so we come to an order which is radically different from that of Galileo and Newton: the order of the undivided whole»[14].

Before Bohm had given the decisive shove to the apparently firm backbone of the classical theory of quantum mechanics, the latter rested on the basis of the Heisenberg's Uncertainty Principle, which states that it is not possible to determine the trajectory of an elementary particle like the electron, since you can't simultaneously know its position and its speed at any given time. Unlike the trajectory

---

[14] D. Bohm, *Universo, mente, materia*, Como: Red edizioni, 1996, p. 176. My translation.

of a bullet in uniform rectilinear motion which moves along a path which is easily predictable by the determinism of Newtonian mechanics, the trajectory of an electron (or any other sub-atomic particle), can't be determined with certainty because it defies all calculations performed in the field of classical Newtonian physics, and the prediction of its movement must inevitably enter into the field of statistical probability.
In this respect, in order to indicate the possible quantum states where the electron may be, it was necessary to introduce a special mathematical function called «the wave function».
Now, the time has come to focus upon the concept of probability: one of the main topics of Quantum mechanics itself concerns the process of measuring. Since measuring the particles in the laboratory perturbs the state of the particles themselves, here's the big, shocking innovation brought by quantum mechanics: the observer affects the observed, the subject influences the object.
At the time it is measured, the particle is located in a specific place, precisely the one which is measured, which has to be different before and after the above measurement, because, as we have seen, it is only the

«external» act of measurement that determines spatiality. The problem therefore, arises by itself: between one measurement and the other, the particle is in an overlapping position of probability waves, which is equivalent to saying that it is potentially present in several places simultaneously.

> «Any attempt at analysis of the «individuality» of atomic processes, intended in the proper sense of classical physics, would be corrupted - conditioned by the quantum of action - by the indelible interaction between the studied atomic objects and the necessary tools of measurement. [...] This point of view, however, does not coincide with an arbitrary renunciation of the analysis of atomic phenomena, rather it would be the expression of a rational synthesis of the complex of experiences within this domain, which extends beyond the limits within which the concept of causality is naturally confined»[15].

The act of measurement essentially promotes a collapse of the particle in a given localised space, however, this location, according to the classical model, is completely random and all

---
[15] *Ibidem.*

other measurements involve further collapses, just as random and mysterious. Erwin Schrödinger offers an interesting possibility of intuitive understanding that sheds light on how the observer acts and interacts with the observed in the field of the collapse of the probability wave function; This is the famous «paradox of Schrödinger's cat».

«You can also create very burlesque situations. A cat is locked up in a steel box with the following infernal machine (which must be guarded against the possibility of being seized directly by the cat): in a Geiger counter there is a tiny amount of radioactive substance, so small that within an hour one of its atoms might disintegrate, but, just as likely, none of them will. If the event occurs, the counter indicates it and triggers a relay of a hammer that breaks a vial full of cyanide. After having left this whole system undisturbed for an hour, one would say that the cat still lives if in the meantime no atom has disintegrated. On the other hand, the first atomic disintegration would have poisoned it. The function $\Psi$ of the entire system leads to state that, within itself, the alive cat and the dead cat are not pure states, but mixed with equal weight»[16].

After a certain period of time the cat has the same probability of being dead as the atom has of decaying. Since, until the time of observation, the atom exists in two overlapping states, the cat is both alive and dead until you open the box, that is until an observation is not made. In short, until the cat is inside the box we do not know if it dies or not, the only thing we know is that there is an overlap of states in which the cat is both alive and dead. The act of opening the box is equal to that of measurement, as it collapses all the possibilities to just one: the measurement destroys the superimposition, transforming *de facto* a possibility into reality.

In essence, an elementary particle has the ability to be placed in different positions and also to be equipped with different amounts of energy in the same instant. As absurd as they may seem, these unusual properties of matter and energy correspond to the reality of the quantum world. Subatomic particles are delocalised in space and motion, so that, between one observation and the other, they

---

[16] N. Bohr, *I quanti e la vita*, Torino: Bollati Boringhieri, 1965, p. 43. My translation.

behave as if they were in several places at once. Only when a delocalised particle is observed during an experiment - which inevitably changes its energy level, quantity of motion and position - it is identified with certain values of its variables amongst the various possible ones.
The German physicist Pascual Jordan sums up:

> «Not only do observations disturb what should be measured, but they create it ... We force an electron to take a definite position... But we ourselves generate the results of measurement»[17].

This means, that what we know about a particle before measurement, is not a reliable piece of information, but only a probability. In other words, it seems that we cannot have a vision of the state of things as they really are; we can know how they are only if we are involved. There is no more separation or dualism between an observer and a given data.

---

[17] Quoted in M. TEODORANI, *Entanglement*, Cesena: Macroedizioni, 2007, p. 9. My translation.

If in 1800 the physicist Thomas Young demonstrated that light was composed of waves, or pure energy; in 1905 Einstein, just as rigorously, showed that light consists of particles, or pure matter. Therefore, both qualities were equally present in light, and the emergence of one over another depended only on the experiment that was being performed.

In the light of current knowledge, Young's experiment is even more determinant and explanatory with regard to the duality of light. Let us consider this procedure in depth.

Suppose you put a monochromatic light source behind a screen on which there are two slits. If you close one of the slits, the light can only pass through one of them, forming, in the reality that it impacts, a ray of light the same shape as the slit. In this case light expresses a corpuscular nature. If instead we keep both slits open, light reveals a wave nature, creating beams of interference that intersect each other (as if, for example, we imagine throwing two stones in a pond: the waves that are created overlap). During Young's days, the necessary knowledge to look into the matter any further was lacking, and everything stopped there, furthermore, the intent of the experiment itself aimed only at demonstrating the wave nature of light. Only later, with the study of interference beams, the fact that light not only had a wave nature, but also a particle one, was noticed.

Not only is the photon actually in two places at the same time, but the photon *seems to know* exactly when to «double» itself.

Everything collides violently with the mechanistic concept of reality, contradicting the principle of causality in favour of *synchronicity*[18].

In the Seventies and Eighties people came to understand that this absurd movement is not only a prerogative of photons, but also of particles of matter, like electrons and neutrons. As stated by Massimo Teodorani:

> «You may also think that particles are kind of "intelligent", since when a particle passes through both slits simultaneously it seems to have a perfect consciousness of the past and the future in order to create the correct figure of interference»[19].

---

[18] Jung and Pauli deepened this issue, about which I refer to my essay *On the concept of synchronicity. Jung between psychoanalysis and quantum physics*, CreateSpace Indipendent Publishing, 2014.

[19] M. TEODORANI, *Entanglement*, cit., p. 16.

In the instant of this creation they solidify their energy in particulate matter.

The Italian scientist summarises the unsettling absurdity of philosophical implications that these discoveries inevitably raise:

> «We can only note that this phenomenon shatters all conceptions of reality that we have built to explain the world in which we live. But quantum mechanics show us that physics is not just meant to describe the world of our sensory experience but also to penetrate into the depths of a world invisible to us, which seems to bear alone the structure of reality as a whole. The notions of reality that we have built in a few centuries of Galilean science are fully consolidated at the level of our psyche and at the level of a common collective consensus, to the point that even Albert Einstein in the last century was to be conditioned by common sense, and to believe throughout his life that quantum mechanics [...] was not a complete theory. According to Einstein, there had to be hidden variables which could be deployed in a random manner, and not in a

synchronic one - as the structure of quantum mechanics actually appeared to be»[20].

During Einstein's days, no one could accept that the general idea of the universe and its functioning was totally inadequate, not even the famous German scientist, who, along with his two associates Rosen and Podolski, asked two simple but determinant questions, which were actually two alternatives (which later, unexpectedly, would even exclude one another): either there are hidden variables, not yet discovered, that impede us to conceive quantum mechanics as a deterministic theory, or the theory of relativity, with its finite limit of the speed of light, is strongly put into discussion.
It's on the bases of these reflections that the famous mind experiment devised by the three scientists, and which took the name of «EPR Paradox» from their initials, was born.
This experiment is particularly important because it was subsequently examined by the Irish physicist John Bell who in 1964 formulated a theorem: the Bell's theorem, indeed. It was then validated by physicist Alain Aspect in 1982 with an experiment that

---
[20] *Ibidem*, pp. 17, 18.

left no doubt about the actual non-locality of particle communication[21].

Bell's theorem is of fundamental importance because it rigorously demonstrated how, from a mathematical point of view, the supposed hidden variables that Einstein hoped to find, are out of the question: quantum mechanics is a completely independent theory which is based on synchronicity and not causality.

As Davies says:

> «In the sixties, the physicist John Bell was able to show that the degree of cooperation between separate and distinct systems may not exceed a certain limit if we assume, as Einstein did, that fragments are well defined before observation. But if you follow quantum theory, this limit does not exist. To solve the problem, there was the need for an experiment»[22].

However, to understand how we got to this point, it's necessary to dwell upon the experiment itself.

---

[21] See A. ASPECT, Introduction to J. S. BELL, *Speakable and Unspeakable in Quantum Mechanics* (Collected Papers on Quantum Philosophy), Cambridge University Press, Cambridge 1987. My translation.

[22] P. DAVIES, *Dio e la nuova fisica*, Milano: Mondadori, 1983, p. 150, My translation.

The EPR paradox takes into consideration a simple elementary particle like the electron with no spin[23]. If such a particle is divided into two parts, one must necessarily have a spin equal to + ½ and the other spin equal to - ½. This is inevitable in order to ensure the law of conservation of the spin, of which the sum must be zero - in the moment in which particles might reconnect. Now, if we launch the two particles at great distances and modify the spin of one of the two, in order to ensure the law of conservation, the other particle must necessarily instantly change its spin. This immediate change, however, if on the one hand ensures that the sum of the spins are to be zero, on the other hand dramatically violates the relativity theory which states that a signal cannot travel faster than the speed of light.

To summarise, the immediate change in the spin of the second particle is, by all means, a

---

[23] The spin is the «intrinsic angular momentum of the electron and of every other particle. In quantum systems it is quantized and can only take fixed values. In the case of electrons, this value is ½ and can take positive and negative values». M. TEODORANI, *Sincronicità. Il legame tra Fisica e Psiche da Pauli e Jung a Chopra*, Cesena: Macroedizioni, 2006, p. 140. My translation.

non-local event totally unexpected by classical physics; in simple words: *classical physics cannot explain this phenomenon.*

Taking everything into account, we may say that un urgent need has arisen for a new physics, a new language, a new vision of the world. And here are the philosophical consequences that twenty-first century science poses and imposes: a science where reality is no longer external to scientists who study it, rather it belongs to them in terms of participation, even creation:

> «The great theoretical physicist John Archibald Wheeler himself, in the latter part of his life (after doing canonical studies on relativity, quantum mechanics, black holes and cosmology) realised that the universe appears to be made of «bits» of information rather than bits of matter and energy. Wheeler then suggested that we live in a «participatory universe» in which we - our act of asking questions about nature - participate in the creation of the observed world»[24].

In such a context, art and science are intertwined, they themselves become

---

[24] *Ibidem*, pp. 34,5.

entangled in a perspective in which creation merges and mixes with study, philosophy with science, being human with being artists, architects of the same perceived reality[25].

In this regard, the references to Hermeticism are plausible and relevant in terms of an integral and interdisciplinary philosophical perspective.

According to Hermetic philosophy, that affirmed the affinity of macrocosm and microcosm, the process of creation of the

---

[25] Nevertheless, the topic is still lively discussed nowadays. It is all about *how* and, especially, *if* the state of an electron can be compared to that of an extended body such as a table or... a human being. This is the fascinating major problem that today's scientists are called to confront themselves with.

Two electrons can certainly be entangled, but what about two human beings? An electron is an elementary particle with virtually no mass and can touch the speed of light, but can a human body do the same? Can it delocalise?

Twenty-first century physics (not that of the EPR paradox, which is early twentieth century) still cannot give a definitive answer to this question, which for now finds feedback and references only in the context of initiatory and mystical traditions – bilocation is a well-documented phenomenon in the history of Eastern and Western mysticism. To this regard, see: E. ZOLLA, *I mistici dell'Occidente*, Milano: Adelphi, 1997; M. ELIADE, *Trattato di storia delle religioni*, Torino: Bollati Boringhieri, 2008. My translation.

world and that of the creation of the self could be associated, a demiurgic act is indeed related to both creative events. In the work *'The Hermetic Tradition'*, Julius Evola asserts: «The process of creation and the process that man through the Arts replenishes himself, follow the same path and have the same meaning»[26].

Creation is possible, it is within the reach of man, although certainly not the common man but a man who is aware, an «enlightened»:

> «In order to appreciate this teaching, we must clearly surpass the idea of creation as a historical fact which ended in the past - spatial and temporal; we must conceive it in terms of a «creative state», metaphysical by its very nature, therefore beyond space and time, beyond both past and future. This is more or less the same concept that some mystics, even Christians, designate as eternal creation. In this way, creation, is an ever present act and consciousness can always retrace it through states, which - according to the «principle of immanence» - are possibilities of its profound nature - its «chaos» - while in cosmogonic myth they

---

[26] J. EVOLA, *La tradizione ermetica*, Roma: Mediterranee, 1971, p. 54. My translation.

are given in the form of symbols, gods and primordial figures and actions»[27].

---

[27] *Ibidem.*

*A religion contradicting science and a science contradicting religion are equally false.*

P. D. Ouspensky

## 3. An interconnected universe: philosophical implications of the new physics.

Contemporary physics is expected to face up to two puzzling matters and the same applies to contemporary philosophers who are called to investigate it: if we interfere with particles, they change. Moreover, when they change they don't do it locally, discrediting all classical physics' concepts with regard to space and time.

Quantum physics, taking cognizance of the absurd behaviour of particles, which is *implied* by the act of the observer, proposes an indissoluble entanglement of subject and object and, therefore, of «mind» and «matter».

As previously explained, what we know prior to the measurement of a particle, is nothing more than a cloud of probability in which the particle might be located.

David Bohm wanted to find out what could lead to this dual and mysterious behaviour. He did so by reformulating the Schrödinger equation, which describes the electron's motion, adding a key parameter: the quantum potential. The success of this insight lies in the fact that Bohm *implicitly* introduced the concept of «synchronicity».

According to the American physicist, particles act in synchronicity by dint of a quantum potential which, invisible and as a matter of fact *noumenously* unknowable, guides and regulates the particles' behaviour from a further, or «parallel», plane. In this explanatory framework, determinism is preserved, but in the light of a process which is utterly inexplicable mechanistically.

> «There is determinism here, but it is very different from Newtonian determinism in which the causes must always precede the effects: in this context, causes and effects coincide and such determinism is not a clockwork mechanism but a sync order of things, much like a living organism in which all its parts act in harmony and where the form is the unifying character of all the intimate elements that make up the universe»[28].

This new worldview implies unavoidable philosophical consequences: «The mathematical method of investigation of physical phenomena is and remains valid, but

---

[28] M. TEODORANI, *Entanglement*, cit., p. 41.

at profound levels, the physicist's mind is also forced to open up to new horizons of thought that are connected in part to Platonic philosophy and in part to Eastern religions»[29]. In this regard, the renowned work *The Tao of Physics* by Fritijof Capra is a classic reference text for the scholar interested in the above-mentioned connection:

> «In recent decades, high-energy diffusion experiments have revealed to us, in the most extraordinary way, the dynamic and ever-changing nature of particles; matter has proved capable of transformation. All particles can be transformed into other particles, they can be created by energy and can disappear into energy. In this context, classic concepts such as «elementary particle», «material substance» or «isolated object» have lost their meaning: the whole universe appears as a dynamic network of energy configurations which cannot be separated»[30].

A connection that once again proposes a harmoniously and synchronously interconnected, and no longer dual, reality:

---

[29] *Ibidem*.
[30] F. CAPRA, *Il Tao della fisica*, Milano: Adelphi, 1999, p. 96.

«A reality that cannot be defined as neither subjective nor objective. The world of matter and that of mind are so intrinsically interconnected as to form a single whole [...]. Yet this concept is not new, but dates back two thousand years ago when the Hindu Tantric tradition of the world postulated a similar philosophy. According to Tantric philosophy, reality is nothing more than an illusion and this illusion is called «veil of maya». The main error that we commit is not perceiving this illusory veil: we perceive ourselves as separate from the world around us. This is a realm where the laws of classical physics no longer apply; it stands for the ultimate goal of physics as well as a major stumbling block: we still can't find the metric, the geometric domain and the mathematical operators able to describe it formally»[31].

Professor Amid Goswami states:

«We all are used to thinking that everything around us is already a thing existing without my input, without my choice. You have to banish that kind of thinking. Instead, you really have to recognise that even the material world around us, the

---

[31] M. TEODORANI, *Bohm...*, cit. p. 36.

chairs, the tables, the rooms, the carpet, time included, all of these are nothing but possible movements of Consciousness. And I'm choosing, moment to moment, out of those movements, to bring my actual experience into manifestation.
This is the only radical thinking that you need to do. However, it is so radical, it is so difficult, because we tend to take the world out there for granted, independent of our experience. It is not. Quantum physics has been so clear about it. Heisenberg himself, co-discoverer of quantum physics said, "Atoms are not things, they're only tendencies."
So instead of thinking of things, you have to think of possibilities.
They're all possibilities of Consciousness»[32].

The inability of describing, part and parcel of this mathematical model, led Bohm himself, in the later part of his life, to take an interest (almost exclusively) in the philosophical implications of his discoveries. His connection with Krishnamurti[33] only aimed to increase

---

[32] From an Interview at "The World Congress of Quantum Medicine", accessed November 30, 2014,: http://www.youtube.com/watch?v=vnzYVJGjjK8

[33] See D. BOHM – J. KRISHNAMURTI, *The Limits of Thought: Discussions between J. Krishnamurti and David Bohm*, Routledge, New York 1999.

this markedly philosophical attitude, unacceptable for the academic world close to him, firmly anchored to the Newtonian view of reality and therefore incapable of accepting a union of philosophy and physics or even physics and mysticism.

According to the bohmian concept, particles do not communicate with each other at super speed, a speed faster than that of light. Simply, «they are never moved», they are never separated, they do not need to move to reach each other, since they are *already reached*. The scientist translates this evidence from the microscopic world to the macroscopic one: individuals themselves are not separate entities, but extensions of one ultimate reality, like many tips of a submerged iceberg, which outwardly appear to be separate but in depth are firmly connected, even if this is invisible - at least to the Newtonian mechanistic eye.

Therefore, if Renaissance philosophers and magicians once talked about *anima mundis*, contemporary physics speaks of universal hologram, zero-point field, quantum potential and implicate order. It is noteworthy to mention that even Nicholas of Cusa spoke of «implicate order», while Bruno himself spoke of *mens insita omnibus* (the mind is in everything) and *mens super omnia* (the mind is over everything).

If, according to the vision of Bohm and Pribram, true reality is at this point only that of the potential quantum field or zero point, the reality that we experience, the empirical world, is nothing more than a hologram, a projection, essentially something entirely illusory: «But the most puzzling feature of quantum potential implies that, in essence, objective reality, despite its apparent solidity, does not exist»[34].

On the scientific horizon *implicated* by these discussions, we can also include so-called «transpersonal psychology»[35], confirming a turmoil of cross-sectional studies that structure, within quantum physics, a highly syncretistic approach. The famous Italian psychiatrist Roberto Assagioli is quite clear about the illusion of reality:

---

[34] M. TEODORANI, *Bohm...*, cit. p. 36.
[35] Born in the United States in the Fifties, whose most celebrated representatives are Abram Maslow, Stanislav Grof and Roberto Assagioli (who partly transcended it founding Psycho-synthesis, «the fifth force» after the fourth that is transpersonal psychology, even if in reality the differences between the fourth and fifth are not as marked).

«The state of consciousness of the ordinary man can be called a dreamy state in a world of illusion: illusions produced by imagination, emotions, mental conceptions and more generally illusion of the external world which our senses perceive as reality. With regard to the external world, modern chemistry and physics have shown that what appears concrete, stable and inert to our senses, on the other hand is a dizzying swirl of infinitesimal elements, energy charges animated by a powerful force. Therefore, matter which appears to our senses and which was conceived by materialistic philosophy, does not exist. Current science has eventually shared the fundamental theoretical construct of India, the ancient spiritual vision according to which everything that appears is *maya*, illusion»[36].

These are the shocking implications with which philosophic thought must now come to terms. The various experiments, tests and demonstrations listed before in this essay, impose a reflection which is anything but brief, rather a whole new systematic organisation of the mechanistic concept of

---

[36] R. ASSAGIOLI, *Lo sviluppo transpersonale*, Roma: Astrolabio, 1988, p. 77.

reality. What quantum physics is revealing will completely subvert man's relationship with himself and with the world around him.
The Dutch physicist Hendrik Casimir, with a famous experiment, managed to discover what is also called the quantum foam, that is the zero-point field. Drawing two plates together, very close to one another, Casimir saw that they were subject to a rather abnormal pressure caused by the energy of the vacuum, which therefore, far from being «empty», contains energy that comes into contact with the remaining quantity of material close to it.
Bohm himself compared the void to the *prana* of Indian philosophy, an energy which is truly significant at the material level, since the two levels are closely interconnected and the material level would be nothing more than filled emptiness or, in Bohm's words, *explicated,* i.e. the invisible made visible:

> «*De facto*, calculations of the quantity known as the "zero-point energy" suggest that each cubic centimetre of empty space contains more energy than all the known matter in the universe, while current models of theoretical physics and cosmology predict that so-called «dark energy» comes directly

from the void and constitutes 73% of the energy produced by the universe»[37].

The energy that comes out of the vacuum, i.e. from an implicated world, is creative energy because it creates the explicate order, that is the material order, and this process - the transmutation from the implicate to the explicate, from the empty to the material - is called, in Bohmian physics, «holomovement». Thought itself, according to the American physicist, is merely explicated energy coming directly from the implicate order, from the empty itself or «pre-space» as Bohm defines it. Such a non-local entity could also coincide with Jung's «collective unconscious», the location of the archetypes from which the synchronicities found in the explicate world originate. The latter were investigated in depth not only by the famous Swiss psychologist but also by the Physics Nobel Prize Wolfgang Pauli, for whom Jung was a therapist for a while, and later a friend and collaborator[38].

Teodorani compares pre-space not only to Jung's collective unconscious, but also to Plato's world of Ideas, and argues that when

---

[37] M. TEODORANI, *David Bohm…*, op. cit., p. 84.
[38] See M. TEODORANI, *Sincronicità*, cit.

scientists or artists have a stroke of genius, or simply an inspiration, they draw that energy from the «hyperuranion» itself:

> «Information resides in a place like Planck's field, which is a kind of absolute reference system of the Universe, an area that unites all creation, and that can only be perceived during moments of consciousness: aesthetic values, the perfection of mathematics, beauty and the most sublime feelings which are part of the Platonic database, that exist on the Planck scale at the time when our microtubules, and the turbines inside them, collapse. [...] Where do you think certain scientists and artists derive their genius from? From their ability to make good use of the moments of consciousness. In essence they are «good antennas», able to access a higher realm that, thinking about it, is very similar both to Jung's collective unconscious and Bohm's implicate order».[39]

Moreover, all of this reveals more than one affinity with the peak experiences of the above mentioned Transpersonal Psychology, culmination experiences in which you can experience high states of consciousness, which can be induced voluntarily, with the

---

[39] M. TEODORANI, *Entanglement*, cit., p. 81.

aid of specific techniques (the Esalen Institute was founded to carry out this type of experiments) or involuntarily, let us think of near-death experiences, whose Dr Elizabeth Kübler Ross was the greatest pioneer[40].

In conclusion, the new paradigm imposed by quantum physics implies consequences at the level of the psyche. To this regard the psychiatrist Pier Maria Bonacina in his work *L'uomo stellare* points out: «If we agree that matter is energy and everything is interrelated, then we need to look at everything with new eyes, with a new mind, new amazement, wonder and enthusiasm. This new look implies a new interest, new freedom and new interpretative models of thought processing, of psychological functions, of personality, of identity»[41].

By and large, investigating the gripping discoveries of quantum mechanics implies implementing dialogue between different languages (scientific, philosophical and esoteric) which in turn will make different world-views coexist, in virtue of a broader

---

[40] To this regard see the celebrated E. KÜBLER ROSS, *On death and dying*, New York: Scribner, 2011.
[41] P. M. BONACINA, *L'uomo stellare*, Firenze: Giampiero Pagnini Editore, 1998, p. 306.

and more comprehensive epistemological perspective.

Award-winning Roger Penrose, Emeritus Professor of mathematics at the University of Oxford and staunch supporter of the quantum reality of consciousness, is peremptory in stating:

> «I feel led to say that clear answers will not come unless you see these worlds interact [the mental world, the physical world and the Platonic - or mathematical - world]. None of these problems will be solved in isolation from all others. I referred to the three worlds and the mysteries that bind them. There is no doubt that there aren't actually three worlds, but only one, the true nature of which we have not even seen for a moment»[42].

In the award-winning documentary film about quantum mechanics, *What the bleep do we know?!*, whose title is quite explicative, it is argued that: «What we know we know, is nothing in comparison to what we know that we don't know. However, that which we know we don't know is nothing in

---

[42] Quoted by M. Teodorani in *Entaglement*, cit. p. 86.

comparison to that of not knowing what we don't know».[43]

We're just at the beginning then, even if well-equipped with an "erudite" ignorance using Nicholas of Cusa's words, an ignorance that is in fact constant and never-ending thirst for knowledge, spasmodic gnosiological longing that surpasses itself in the very act of emerging - this specific act is its primal impulse matrix as crossing the limits is its main faculty.

According to a Sufi proverb man is an empty cup and this is good because an empty cup can't but be filled; that is the meaning of his existence, he has no other purpose besides filling.

We can look to quantum mechanics as an effective opportunity to fill the (alleged) *void of knowledge* that constantly tries to challenge man, compromising the *probability of being himself*.

---

[43] See W. ARNTZ - B. CHASSE - M. VICENTE, *What the bleep do we know?!* New York: Samuel Goldwyn Films, 2004

# REFERENCES

R. ASSAGIOLI, *Lo sviluppo transpersonale*, Astrolabio, Roma 1988.

J. S. BELL, *Speakable and Unspeakable in Quantum Mechanics (Collected Papers on Quantum Philosophy)*, Cambridge University Press, Cambridge 1987.

M. BAIGENT – R. LEIGH, *L'elisir e la pietra. La grande storia della magia*, tr. it. di S. Lalia, Il saggiatore, Milano 2003.

D. BOHM – J. KRISHNAMURTI, *The Limits of Thought: Discussions between J. Krishnamurti and David Bohm*, Routledge, New York 1999.

D. BOHM, *Universo, mente, materia*, Red edizioni, Como, 1996.

- *On creativity*, Routledge, New York 1996.

N. BOHR, *I quanti e la vita*, Bollati Boringhieri, tr. it. di Pyoung. Gulmanelli, Torino, 1965.

P. M. BONACINA, *L'uomo stellare*, Giampiero Pagnini Editore, Firenze 1998.

F. CAPRA, *Il Tao della fisica*, Adelphi, Milano, 1999.

P. DAVIES, *Dio e la nuova fisica*, tr. it. di M. Paggi, Mondadori, Milano, 1983.

M. DONA', *Magia e filosofia*, Bompiani, Milano 2004.

L. GIULIODORI, *Sul concetto di sincronicità: Jung tra psicanalisi e quantismo* in «Schegge di

Filosofia Contemporanea», Decomporre, Gaeta 2014.

P. GUGGISBERG NOCELLI, *La via della psicosintesi*, L'Uomo Edizioni, Firenze 2011.

J. EVOLA, *La tradizione ermetica*, Mediterranee, Roma 1971.

E. KÜBLER ROSS, *La morte e la vita dopo la morte*, tr. it. di M. F. Sanguinetti, Mediterranee, Roma 1991.

G. LA PORTA, *Giordano Bruno. Vita e avventure di un pericoloso maestro del pensiero*, Bompiani, Bologna 2001.

K. PRIBRAM, *I linguaggi del cervello*, Franco Angeli, Roma 1980.

F. PULLIA, *Giordano Bruno tra Oriente e Occidente*, in "Testimone dell'infinito. Giordano Bruno 1600-2000", Atti del Convegno Perugia-Terni, Ali&No Editrice, Perugia 2004.

M. SGALAMBRO, *Anatol*, Adelphi, Milano 1990.

M. TEODORANI, *Sincronicità. Il legame tra Fisica e Psiche da Pauli e Jung a Chopra*, Macroedizioni, Cesena 2006.

- *David Bohm. La fisica dell'infinito*, Macroedizioni, Cesena 2006.

- *Entanglement*, Macroedizioni, Cesena 2007.

F. YATES, *Giordano Bruno and the Hermetic Tradition*, Routledge and Kegan Paul, London 1964.

E. ZOLLA, *I mistici dell'Occidente*, Adelphi, Milano 1997; M. ELIADE, *Trattato di storia delle religioni*, Bollati Boringhieri, Torino 2008.

- *Auree*, Marsilio, Venezia 1995.

- *Verità segrete esposte in evidenza*, Marsilio, Venezia 1990.

www.luciogiuliodori.net

www.luciogiuliodori.net

www.ingramcontent.com/pod-product-compliance
Lightning Source LLC
Chambersburg PA
CBHW071806170526
45167CB00003B/1194